Grow the Best Corn

Nancy Bubel

CONTENTS

Introduction

Corn is a New World plant, native to parts of Mexico and Central and South America. When Columbus landed in the New World in 1492, native American Indians had been growing popcorn, sweet corn, dent corn, and flint corn for hundreds of years. They had developed these distinct varieties by repeatedly selecting the best ears of each type and saving and planting the seed.

Sweet corn is a gardener's vegetable, one of the most eagerly awaited summer crops. By growing your own, you can have it at its best: sweet, tender, juicy kernels, five minutes from the patch. Fresh sweet corn is so good that it needs no sauces or fancy recipes; simply steam and serve. Many gardeners like to eat raw corn while husking it. Corn is a vigorous plant that responds to generous fertilizing, so it is satisfying to grow.

Corn's best features are its sweet, delicate flavor (sweet corn) and its versatility (popcorn, flint, and dent corn). On the negative side, corn takes more growing space than many other vegetables, uses up a lot of soil nitrogen, and — if grown in large plots — exposes the soil to erosion. There are ways of getting around these difficulties, though, as we will show you in this bulletin.

Kinds of Corn You Can Grow

Usually, selecting a vegetable variety to grow is a pretty easy choice — mainly because you do not have that many varieties to choose from. But if you have ever counted the varieties of corn offered in most seed catalogs, you know that selecting a variety of corn to grow involves considering a lot of different factors. (On the last pages of this bulletin there are more than three dozen recommended varieties — and that's just a sampling of what is available!)

To select a variety of corn to grow, first decide whether you want to grow sweet corn, popcorn, flint corn, dent corn, flour corn, super-sweet corn, or high-lysine corn. Then you will want to choose between hybrid and open-pollinated varieties.

Types of Corn

Let's consider, first, the characteristics of each of the different kinds of corn you might want to grow, again with a bow to the native Americans, who, without tools, or books, or an understanding of the principles of botany, managed to divide the corn family into five distinct groups: sweet corn, popcorn, flint corn, dent corn, and flour corn. Recently, plant breeders have given us new types of corn to consider, including super-sweet varieties and high-lysine corn.

Sweet Corn. This demanding variety is so good that it is worth extra soil feeding and careful timing to pluck its perfection from the corn patch. Some early sweet corn is ready as soon as fifty-five days from planting, making it possible, in some areas, to follow the corn with a fall vegetable crop or a soil-improving cover crop. Sweet corn thrives on hot weather, plenty of soil lime and nitrogen, good drainage, and abundant moisture.

Kernels of sweet corn contain a relatively small amount of starch and a large amount of water containing sugar in solution. When dry, the kernels shrink and wrinkle. The characteristic sweetness and tenderness of sweet corn are actually caused by a genetic defect that keeps the starch grains in the kernel few and small and prevents the conversion of soluble sugars into more starch.

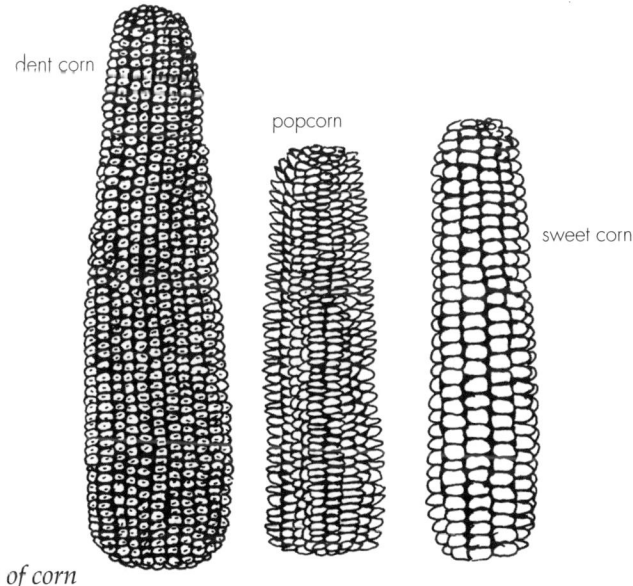

dent corn

popcorn

sweet corn

Types of corn

Sweet corn is generally less robust in germination and growth than dent corn, flint corn, or popcorn. Also, it is somewhat more demanding of rich soil, and often less disease-resistant.

Popcorn. Ears of popcorn are shorter than those of most kinds of sweet corn, and the kernels are smaller. Each kernel contains a small central core of soft starch around the germ, surrounded by a larger outer layer of hard starch. The kernels pop when heat expands the moisture in the center of the kernel, and the resulting steam bursts the hard, starchy coating.

Like sweet corn, popcorn does well in hot weather and well-drained, limed soil; but it will thrive on somewhat leaner soil and less moisture than sweet corn. Most popcorn varieties need 90 to 100 days to mature. Some, like strawberry popcorn, are also decorative and frequently grown for roadside stands.

Flint Corn. You do not hear much about flint corn, but it is a useful and delicious staple corn, good for year-round storage. The kernels are hard and smooth, not shriveled like sweet corn or indented like dent corn. The hard kernels can be ground to produce a sweet, nutritious meal for use in making muffins and cornmeal crackers, mush, and bread. Flint corn needs a rather long growing season — about 110 to 120 days — but it is more tolerant of cool weather and prolonged dampness than sweet corn. Flint corn can be found in varieties colored red, blue, black, orange, and mahogany.

Dent Corn. Dent corn, also called field corn, has a dimple at the tip of each kernel, caused by the rapid drying of the soft starch at the center of the kernel. Immature field corn can be sweet, and it is perfectly good to eat — although it is neither as tender nor as sweet as sweet corn. Both the plant and the cob are large. You might want to raise small plots of dent corn for feed if you keep animals. The kernels contain more soft starch than flint corn. They can be ground in a hand or powdered mill.

Flour Corn. Varicolored like flint corn, flour corn has much more soft starch in the kernel. It grinds to a fine flour, rather than the coarse meal produced by flint and dent corn.

Super-Sweet Corn. These varieties sound like the answer to a gardener's dream. Most of the new extra-sweet corn varieties are not only sweeter than standard corn to start with, they also hold their sugar content for a longer time after picking, because conversion of their sugar to starch takes place more slowly. It is not true, though, that they increase in sweetness after picking, only that their sugar content twenty-four hours after picking is higher than that of day-old regular sweet corn, a quality that has made these varieties especially useful for sale at roadside stands.

This extra sweetness has been achieved by tinkering with the genes of standard sweet corn. As is usual in nature, though, something's got to give. In this case, some plants tend to have spotty germination, lower vigor, and sometimes ears incompletely filled with kernels. Breeders are working on these problems, and at least one new extra-sweet-corn type shows improvement in germination and seedling vigor.

The other limitation that has discouraged many backyard gardeners from growing the super-sweets is the fact that their quality suffers if the plants receive pollen from any other kind of corn plants, even different varieties of sweet corn.

The latest improvement in super-sweet corn has been the development of the everlasting heritage varieties. These extra-sweet ears are not affected by cross-pollination. Because they do not require isolation to ensure sweetness, these new varieties should be easier to fit into a backyard garden patch.

High-Lysine Corn. Another recent development, high-lysine corn, looks promising as a source of better-quality vegetable protein. Amino acids are the building blocks of protein. To be complete, a protein food must contain all eight of the amino acids that our bodies cannot make. Most vegetable protein is incomplete; it contains some, but not all, of the essential amino acids. To get all the necessary amino acids, you must eat two or more complementary vegetables at any one meal (corn eaten with beans will give you a complete protein, for example). Regular corn is deficient in the amino acid lysine. The new high-lysine corn contains about the same amount of protein as regular corn, but the protein is more complete, and the corn more nourishing, since the lysine content has been increased by as much as 69 percent. It is commonly grown as field corn for animal feed, but it can be eaten by people, too.

Choosing between Hybrid and Open-Pollinated Varieties

Because its male and female blossoms (tassels and silks) are so distinct and far apart, corn is an especially easy plant to crossbreed. Hybrid corn seed is the result of mating two different strains of corn which have been especially selected for the inherited characteristics they will pass on to the next generation. This first generation, which scientists call F_1, is sometimes an outstanding improvement over both parents — a phenomenon known as hybrid vigor. Many of the new hybrid corns are really super: high-yielding, flavorful, large, and early to mature.

Seed obtained from hybrid crops produce plants that do not duplicate the qualitities of the parent plants. Rather than resembling the parents, plants grown from such seed will revert to the less desirable ancestral forms of the plant, which were used as stepping stones to get to the right combination of genes that succeeded in producing outstanding first-generation plants. Therefore, seed from hybrid corn cannot be saved for replanting, but must be purchased every year.

Hybrid plants, with their more highly controlled genetic makeup, tend to ripen uniformly within a narrow time span. This narrow

Corn Anatomy

We will be referring to various parts of the corn plant in this bulletin, so here is a quick rundown on those parts and their functions.

Brace roots: An arching crownlike group of aboveground roots originating low on the cornstalk and growing down into the ground. Corn grows so fast and is so tall that it needs an extensive root system.

Ear: A cylindrical arrangement of corn kernels growing on a cob. In a good year, most plants produce two ears.

Endosprem: Material stored in the seed for the purpose of nourishing the embryo plant. In corn, the endosperm is the part we want to eat.

Husk: The protective leafy membrane that encloses each ear of corn.

Kernel: An individual grain of corn. It's not exactly a seed, but rather a ripened plant ovary containing a single seed.

Leaves: Serve to produce food for the fast-growing corn plant and also to channel rain and dew toward the thirsty roots.

Silks: The pollen-receptive female part of the corn plant. Each thread of silk leads to an ovary, and each one that is pollinated grows into a kernel of corn.

Tassel: The pollen-bearing male flower of the corn plant.

Tillers: Extra stalks growing from the base of the plant, usually one on each side. They help to nourish the plant and should not be removed.

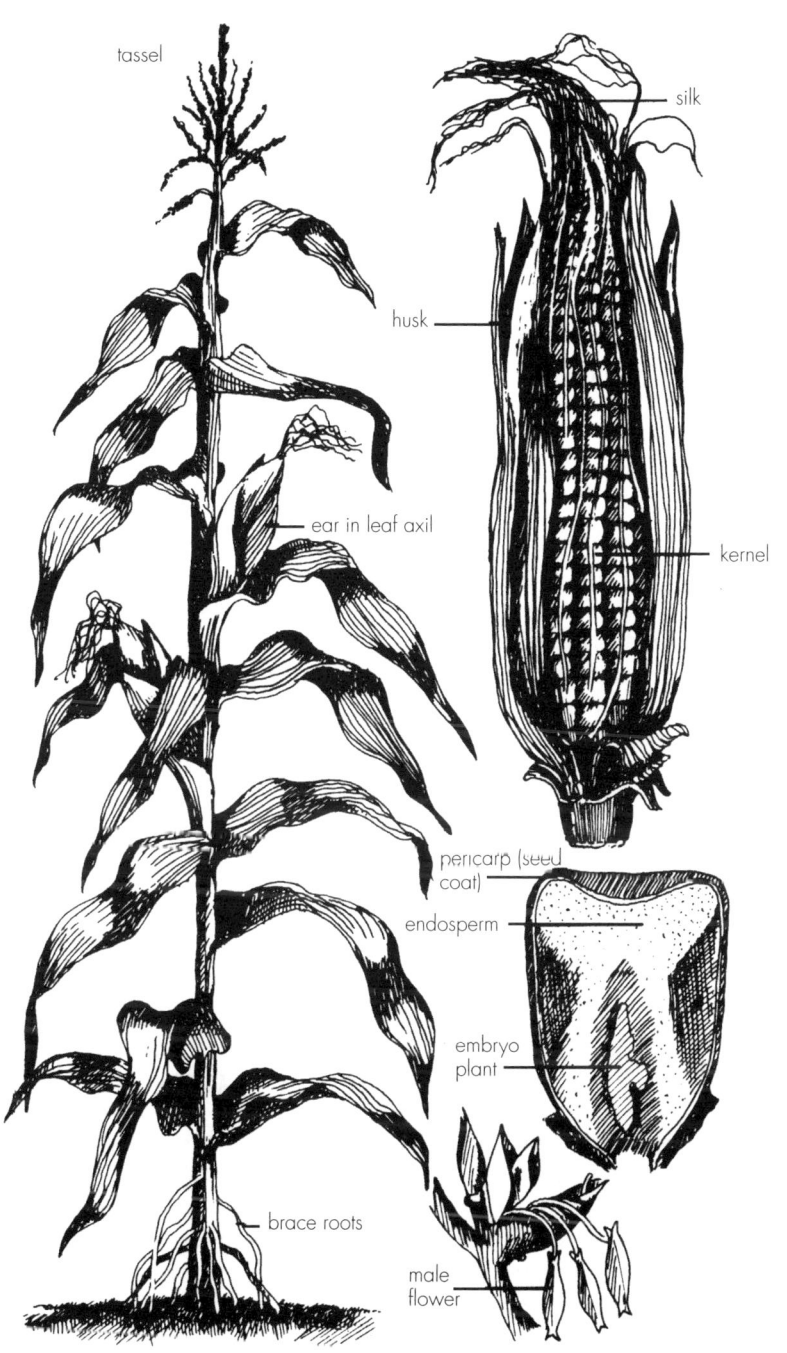

tassel

silk

husk

ear in leaf axil

kernel

pericarp (seed coat)

endosperm

embryo plant

brace roots

male flower

genetic base caused problems for American farmers in 1970, when most of the field corn crop was grown from seed having a common ancestor. These hybrids proved to be vulnerable to a newly resistant form of corn blight, which drastically reduced harvests for that year — a danger signal that has prompted new studies on the importance of genetic diversity.

Open-pollinated sweet corn, the product of uncontrolled pollination, includes such old favorites as Golden Bantam, Country Gentleman, Stowell's Evergreen, and Black Mexican. These varieties, with their more varied genetic backgrounds, ripen over a longer period of time than hybrids. This can be an advantage for the backyard gardener who would rather enjoy a prolonged harvest than freeze corn by the bushel one week and have none the next week. Other advantages of growing open-pollinated corn include the possibility of saving seed for next year's crop and the chance to develop gradually a strain of corn that is well adapted to your local growing conditions.

Very little serious work has been done on improving open-pollinated sweet varieties since the introduction of hybrid sweet corn in the 1930s. Many gardeners have been won over to the excellent taste and tenderness of the new hybrids; but those who remain loyal to the older open-pollinated corn usually have strong feelings about its quality.

A good many of the old favorite varieties of open-pollinated popcorn including Strawberry, South American, and Rhodes Yellow, are still available. Each seed house seems to have its own popcorn specialties, both hybrid and open-pollinated. Most varieties of flint and flour corn that you can buy are open-pollinated. Open-pollinated dent corn is sold by several seed companies, and, of course, many hybrid varieties are widely sold.

Good Soil for Corn

Corn is a member of the grass family. If you think of it as a great big, fast-growing, tasseled grass, you will appreciate what a hearty appetite it has. To support the rapid development of its stout stalk, long stiff leaves, and heavy ears, corn needs a ready supply of plant food, especially nitrogen, which fosters leaf growth. Most gardeners sensibly plant corn in the richest soil on the place.

Enriching the Soil

Enrich the soil well in advance of planting. If possible, plow under a one-inch layer of manure the preceding fall. You can also grow a green manure crop, such as buckwheat, oats, clover, rye, winter wheat, or vetch. In the spring, before planting corn, turn this cover crop under.

Humus. Humus-rich soil contains more air spaces, making it easier for roots to grow. Even more important in the case of corn is the fact that humus acts like a sponge, holding from five to ten times its own weight in water, and gradually releasing it as needed. Nutrients must be dissolved in water in order for plant roots to absorb them; so you can see the value of humus as a reservoir for the rich broth that has percolated through the soil after rain.

Green manure crops and compost build humus. Many gardeners bury their summer household garbage in the center of their corn rows. Any organic materials that you can compost and add to the soil will benefit your corn.

Lime. Corn needs a good supply of lime. The average recommended lime application is a half ton per acre every five years, or as needed. For the home garden, one three-gallon bucketful applied to a plot measuring 50 feet by 20 feet (1000 square feet) every three to four years should be sufficient.

Crop Rotation

Rotate crops to prevent one-sided depletion of soil nutrients. Corn needs a lot of nitrogen. Peas, clover, soybeans, and other legumes have bacteria living in nodules on their roots that add nitrogen to the soil, as much as 100 pounds per acre in the case of alfalfa. To make the best use of this free fertilizer, plant corn right after a legume crop, which can be either a crop of garden vegetables (peas, beans, limas, soybeans) or a hay or green manure crop like alfalfa, clover, or vetch.

Some gardens have special soil problems, such as a hardpan layer of soil, six to twenty-four inches deep. The hardpan is so dense that plant roots cannot penetrate it to get the nutrients they need from the subsoil. If this is your problem, you could substitute a planting of cow-horn turnips (sown at the rate of two pounds per square acre) for one of the other recommended cover crops. The deep roots of the turnips help to break up the subsoil, and leave drainage spaces when they are turned under to decay and add organic material to the soil. Alfalfa, which is slow to get established the first year and needs well-limed soil, is another deep-rooted cover crop sometimes used for this purpose.

Even if you do not find it possible to plant all the recommended cover crops, you should still hopscotch your corn plantings around the garden, alternating corn with root crops, leafy crops, legumes, and fruit-bearers (like tomatoes and strawberries), in order to keep the soil in balance.

Prevent Erosion

Corn is hard on the soil in two ways: it uses up many nutrients, and it leaves the soil exposed to erosion from wind and rain. A closely followed cover-cropping system helps to prevent erosion by keeping the soil covered with soil-holding plants. There is hardly ever any bare soil exposed to the elements under this system. In a very small corn patch, you can mulch during the summer and leave the mulch and corn stubble over winter to catch snow and blown leaves and protect the soil. If your garden is on a hill, run your corn rows around the hill rather than up and down, so that rain will not run in long paths, which tend to cut into the soil and wash it away.

Planting Corn

Hungry as we are for that first taste of fresh sweet corn, planting it too early will result only in wasted seed and row space. Some like to gamble on a small extra-early planting, and there is nothing wrong with that — as long as you know what you are up against.

Timing

Timing corn plantings late enough to be safe, yet early enough to satisfy the restless green thumb and hungry palate — that is one of the arts of gardening. The native Americans made their first planting when the new oak leaves in their neighborhood were the size of squirrels' ears, and that is still a good guide. It is usually safe to plant

Last Expected Frost Date in the Spring

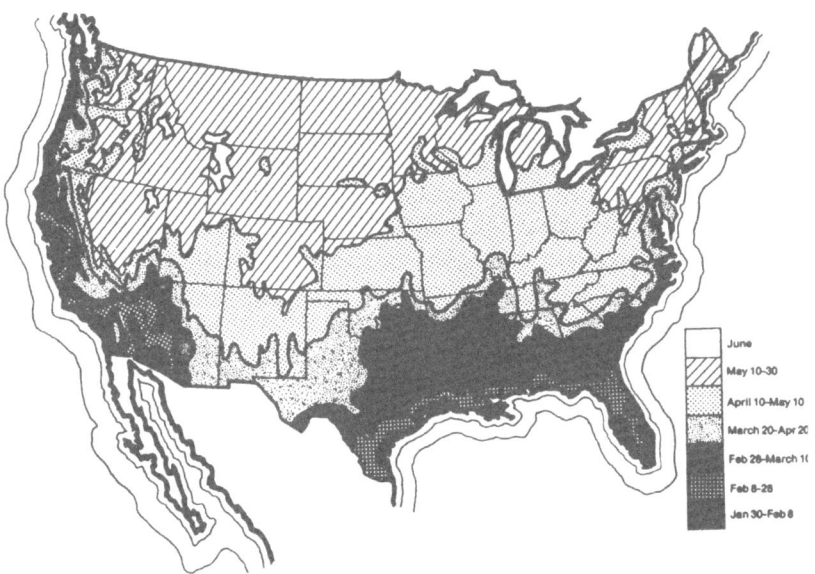

	June
⁄⁄⁄	May 10-30
	April 10-May 10
	March 20-Apr 20
	Feb 28-March 10
	Feb 8-28
	Jan 30-Feb 8

early corn about a week before the date of the last expected frost in your area.

Corn seed, even when treated with a fungicide like Captan, as most seeds are, will not germinate where the soil temperature is below 50°F — the absolute minimum temperature for germination. For early plantings, when the soil temperature ranges between 50 and 60°F, treated seed will germinate much more completely than untreated seed, especially if the treated seed you use is one of the varieties that has been bred to withstand cold soil, like Polarvee or Early Sunglow. When the soil is too cold, seed rots before it can germinate. Rotting seed attracts corn maggots, giving you two problems: an uneven stand and an insect infestation. Early-planted seeds should not be in touch with fresh manure, or they will be more likely to rot than to germinate while soil is still cool.

For a continuous harvest, make repeated plantings — either every two weeks, using small blocks of space, or every three weeks with larger blocks for better pollination. Since early, midseason, and late sweet corns have successive maturation times, many gardeners plant them at about the same time. There is no point in putting midseason or late varieties in the ground extra early; later plantings almost always catch up to the too-early ones. Corn plantings made in midsummer to ripen in early fall can be sown in the pea patch right after the removal of the pea crop. Use an early, quick-maturing variety for these late plantings because growth is slower in September when days grow shorter.

Spacing

Spacing is determined, to some extent, by the kind of corn you plant. In general, corn rows are made thirty to forty inches wide. Choose the narrower spacing for short plants and hand cultivation, the wider to accommodate machines and extra tall plants, to lave room for broadcasting fall cover crops between rows, and when interplanting squash and beans. Early corn varieties grow fairly short plants: four to six feet tall, with relatively small ears. The midseason corns are taller: six to eight feet. Late corn is even taller and more robust: seven to ten feet, with long ears. Many varieties of popcorn have shorter stalks: three to five feet. These can be spaced somewhat closer than the largest corns — rows two to three feet apart, with dwarf plants four to six inches apart. Larger popcorn varieties should be spaced eight to ten inches apart. Plant popcorn about the time of your last frost.

In the row, the usual practice is to sow seeds of sweet, dent, and flint corn 4 inches apart and thin to 8 to 12 inches apart, using the wider spacing for late corn and sowing more thickly (but thinning the same) when planting early corn in cold soil. Cover seed no more than 1 inch deep for early planting, up to 1½ to 2 inches deep later in the season. Some gardeners prefer the hill system; they plant three seeds close together every 30 inches, and they do not thin the plants.

CORN

INTERCROP WITH PUMPKINS OR SQUASH

CORN

CORN

INTERCROP WITH BLACK-EYED PEAS

CORN

EARLY CABBAGES

FOLLOWED BY POTATOES

SWISS CHARD OR CUCUMBERS

TOMATOES

INTERCROP WITH LETTUCE

TOMATOES

INTERCROP WITH GARLIC

PEAS

FOLLOWED BY PEPPERS

PEAS

FOLLOWED BY EGGPLANT

SPINACH

FOLLOWED BY BUSH BEANS

ONIONS

INTERCROP WITH LETTUCE

BEETS

INTERCROP WITH KOHLRABI

CARROTS

FOLLOWED BY RADISHES

MARJORAM

BASIL

MARIGOLDS

BROCCOLI

CHIVES, SHALLOTS

MARIGOLDS

A plan for a small garden based on both intercropping and succession plantings. The corn is interplanted with squash and shell beans. Note that the corn is located on the north end of the garden so that it will not shade other crops. It was planted in a four-row block to ensure pollination. (adapted from Success with Small Food Gardens *by Louise Riotte, Garden Way Publishing.)*

Corn in Small Gardens

Most gardeners who have the space grow large patches of corn to have enough for freezing as well as eating fresh. Where space is limited, though, it is still possible to grow corn for immediate use, which is when it is at its best. This can be done in two ways.

- Devote a small square of the garden, say 10 feet by 10 feet, to a single planting of corn. For greater production, this small plot should be generously mulched, irrigated, and given side-dressings of manure tea (see page 17). Space your plants 8 to 12 inches apart in the row. You can, however, reduce the separation between rows to 2½ feet, if you are able to hand-cultivate this small plot.

or

- Plant a solid block of corn — with plants spaced ten inches apart each way — in a raised bed, which can be worked from all sides. Popular bed sizes are three feet wide and six to ten feet long.

Interplanting. Squash and climbing beans can be interplanted with corn, an old Indian trick that makes good use of garden space. You can plant squash and pumpkins along with corn in the end row, and the vines will ramble into the corn patch and mulch the ground. Or plant a bean seed or two beside each young corn seedling left after thinning the row. The beans return nitrogen to the soil, and the corn stalks support the bean vines. Rows should be more widely spaced to prevent shading of one row by another.

Pollination

Corn is pollinated by the wind, which blows pollen from the tassels of one plant to the silks of another. For well-filled ears with no missing kernels, each silk must receive a grain of pollen. So naturally, you will want to lay out your corn rows so that the wind will carry pollen from corn plant to corn plant, not from corn to other garden plants. The best way to ensure complete pollination is to plant corn in blocks at least four rows wide.

How Cross-Pollination Affects Corn

With most vegetables, cross-pollination between varieties does not show up until the resulting seeds are planted and growing. Corn is different, though. Each pollen grain releases two sperm. One sperm fertilizes the plant embryo and the other joins the nuclei in the endosperm to determine the nature of the stored material in the seed. Because of this unique double fertilization, the kernel that forms on an ear or corn this year can be affected by this season's crossing. If the silks on an ear of sweet corn receive pollen from dent corn tassels, for example, the kernels that grow from these dent corn pollen grains will be starchy and tough like dent corn, not sweet and tender, like sweet corn. Popcorn, flint corn, and flour corn may also cross with sweet corn. If only a few grains of odd pollen are involved, the quality of the ear suffers little, if any; but if half the silks of an ear or sweet corn are fertilized by pollen from popcorn tassels, then you will have an ear of corn that is not so sweet, but it will not pop too well either. You might have noticed, in growing white or bicolor corn next to yellow corn, that you get a few maverick white kernels in the yellow ears if the wind shifts during tasseling. Crossing between regular sweet corn varieties does not usually affect flavor.

Buildings, hedges, or several rows of tall plants like sunflowers can effectively bock the transfer of pollen and reduce the necessary isolation distance between corn varieties. Broom corn, which is actually a sorghum, looks like a corn relative but is not related closely enough to cross, and may be planted close to all kinds of regular corn (*Zea mays*) without fear of crossing.

Saving Seed. If you want to plant corn next year from seed you save this year (more about that on page 25), the plants from which you save seeds should be pollinated only by their own variety, or the next crop will not come true. If you want to maintain a strain of Golden Bantam, then the seed you save must be pollinated only by Golden Bantam corn plants. So you will have to plan your planting schedule so that no other corn crop within 500 to 1300 feet is in tassel at the same time as your seed crop. You can guarantee this by planting early corn next to late corn, for example, because the early corn will have shed its pollen by the time the late corn is in tassel.

There should be at least two weeks difference between maturity dates of different varieties to eliminate all chance of cross-pollination. Plants raised for seed must be more strictly isolated from other tasseling varieties than those raised for food. The reason for this is that a few odd kernels on a cob may not affect its eating quality, but if you plant them, these kernels (seeds) will not grow into the kind of corn you are aiming for.

Super-Sweet Corn. Super-sweet corn (except for the everlasting heritage varieties) loses sweetness when pollinated by other varieties, even other kinds of sweet corn. Super-sweet varieties that are adversely affected by cross-pollination must be planted 350 to 500 feet away from any other tasseling corn. Even at this distance, an occasional odd kernel may develop from alien pollen, but there should be too few of these to affect quality.

Care of the Growing Crop

Care during the growing season should help the plants to make the most of their natural advantage of efficient metabolism. Since corn needs all the light it can get, you will want to avoid planting other tall plants nearby that might shade the young corn plants on their way up. And you will want to make sure that each plant has every opportunity to make good use of the rich soil you have provided. Thin the seedlings to their proper spacing — eight to twelve inches apart. Crowded corn plants shade each other and compete for soil nutrients. Leave the tillers (those extra stalks growing from the base of the plant) on the plant. Although their function is not completely understood, it is generally agreed that corn grows better if the tillers are not removed.

Cultivation

When the plants are small, weeds should be eliminated — or at least kept under control, because they use up soil nutrients, and the fast-growing weeds can shade out young corn seedlings. When chopped down young and tender, weeds contribute a certain amount of green manure. If weeds are hoed down until the plant is knee-high, it is usually not necessary to do much cultivating after that. Some gardeners hill up the rows — draw soil toward the plants from the middle of the rows — when plants are knee-high. This helps to encourage strong root formation. Once they have reached that height, corn plants have extensive roots, which could be damaged by hoeing too close to the plant.

When growing small blocks of corn, some gardeners like to mulch to discourage weeds and retain soil moisture, and this is a good practice to follow as long as you wait until the soil is thoroughly warm before spreading mulch.

Water

Irrigation is not always possible — many corn patches are far from a hose outlet — but in a dry season it can boost your corn yield. Plant food taken from the soil can be absorbed only in solution. And, of course, most plant tissues are composed largely of water. Furthermore, a full-grown corn plant can lose as much as a gallon of water a day in hot summer weather by evaporation through its leaves. If weather is dry, and if you are able to water your corn plants, watering is most effective at the time of tasseling and when the kernels are forming. If you do irrigate your plants, soak the soil at least four inches deep. Surface dampness only encourage the development of shallow roots, which quickly die when the soil dries again.

Fertilizing

For really spectacular corn, try side-dressing the plants twice during the growing season with liquid plant food, such as diluted fish emulsion (available in most garden stores) or manure tea.

To make manure tea, put several shovelfuls of manure into a burlap or cloth sack and suspend this large "tea bag" in a large bucket of water — at least five gallons. Cover the brew and let it steep for several days. Then pour out one to two quarts of the murky water into a three-gallon bucket or watering can. Dilute the concentrated tea with water until it is the color of weak beverage tea. Pour it around the base of each plant. Refill the large bucket with water and let it steep again.

Soak the soil around each plant with the manure tea or fish emulsion once just after the final thinning, when the seedlings are about eight inches high, and again when the plant is knee-high. Many gardeners think such side-dressing is worth their while, to give the plants a boost and to maintain soil fertility. In good rich soil, though, you should get a good corn crop without this extra step.

Troubleshooting

There is not always time to plant a second plot of corn, so damage in the patch can be pretty upsetting. We hope none of the following problems will be yours.

Insects

In the case of insect infestations, an ounce of prevention is always better than pounds of cure. Take measures early, before a full-scale invasion.

Corn Rootworm. The larva of the striped cucumber beetle, the corn rootworm, gnaws on roots and sometimes weakens the plant, but in general does less serious damage than other common corn pests. Rotating corn with soybeans helps to control the rootworm.

Corn Earworm. This pest is the larva of a moth that lays about 1000 eggs in her twelve days of life. Earworms feed on the silks first, turning in the kernels when the silks dry after pollination. In a bad infestation, ears may be heavily damaged, but light damage to the tips is more common. Probably worst of all is that not many families enjoy eating "high-protein" corn with bug larvae on it. (In mild cases, larvae can be removed when the corn is husked.) Earworm damage is more severe in the South. There are three things you can do to defend your corn ears against earworms.

- Squirt mineral oil on the silks when they are turning brown. Never apply oil while silks are still all green, because it will interfere with pollination.
- Cut off the silks close to the ear and discard the egg-con- taining silks. Repeat every four days.
- Find out when the earworm emerges in your area (call your county extension agent) and, if possible, time your corn plantings to mature be- tween the first and second summer earworm hatches.

Corn Earworm (USDA photograph)

Spray mineral oil on the browning corn silks to foil the corn earworms. Make sure you spray after pollination has taken place.

Corn Borer. More destructive yet, the borer hides on one of the many different kinds of plants it can feed on, and moves to the corn patch when the stalks begin to mature. Watch for eggs on the backs of the leaves. Eggs of the European corn borer are white and appear in flat clusters on the leaf undersides. This pest is most active in the North. Eggs of the Southern corn borer (found in the Southern states) vary in color from reddish brown to white, and they appear in groups of fifteen to twenty overlapping scalelike forms on the backs of leaves. A bent-over tassel is often a sign of borer infestation. To control the borer, try the following.

- With your fingernail, split the stalk below the entrance hole of the bug and remove the offending larva.
- Encourage lady bugs, which eat borer eggs.
- Apply a spray of nontoxic *Bacillus thuringiensis* (a caterpillar disease, sold commercially as Dipel or Thuricide), concentrating on the top whorls and tassels of the plant.
- Shred, bury, or compost corn stalks at the end of each growing season so the borer cannot overwinter in them.

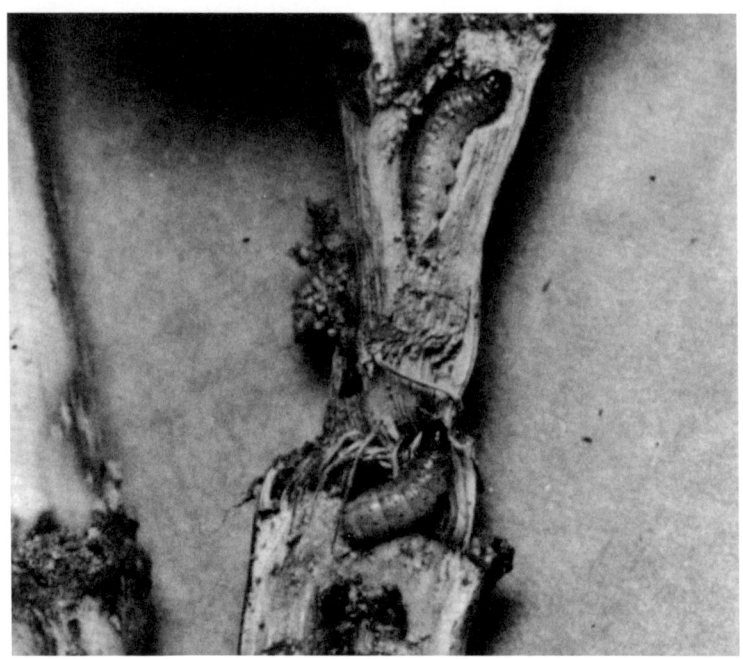

European Corn Borer (USDA photograph)

Diseases

Although corn is vulnerable to a number of diseases, only a few need concern you in a small patch.

Bacterial Wilt. Usually more serious after a mild winter, wilt is caused by bacteria spread by the corn flea beetle. Affected plants may be dwarfed, with early fading tassels; and in late spring, the young plants show yellow patches running parallel to the leaf veins. The best defense is to plant resistant varieties.

Wilt-resistant sweet corns include Golden Cross Bantam, Comanche, and Bellringer. Comet is a resistant white variety and Bi-Queen, a bicolor. Among hybrids, taller and later varieties tend to be more resistant than early, short strains.

Corn Smut. A gall-like fungus growth, corn smut appears as a greyish white membrane filled with black dustlike spores. It is found on the ear. Infestations tend to be worse in hot dry weather. Control the spread of the fungus by breaking off and burning affected ears.

Corn Smut (USDA photograph)

Northern Leaf Blight. This disease hits hardest in humid areas with heavy dew and abundant rain. Plants growing in soil overly rich in nitrogen and deficient in potassium are most vulnerable. Tan or greyish green elliptical patches appear first on the lower leaves. In severe cases, leaves die; then the entire plant dies from lack of nourishment. Blight affecting hybrid dent corn caused alarming crop losses in 1972, leading, again, to an increased awareness of the importance of genetic diversity. Rotation of crops and turning under corn crop residues help to control leaf blight, which is seldom a serious problem in the home garden.

Animals

Raccoons, woodchucks, and deer are probably the worst four-footed sweet-corn thieves. They have an uncanny way of knowing just when the ears have reached their prime, too. Such late damage seldom gives you time to replant. It's really maddening.

Fencing keeps these pests out, but it is an expensive solution. To keep deer out, a fence should be six to eight feet high. Coons and woodchucks respect a four-foot-high fence, although the chucks can dig under if they are ambitious, so you might have to sink the wire fencing as few inches underground for them. A somewhat wobbly fence helps to discourage these critters from climbing over. A strand or two of electric wire around the patch is pretty good corn insurance and easier to install than woven fencing.

Other raccoon-spooking tactics include playing rock music in the patch (leave a battery-operated radio on at night). Walking barefoot around the patch three times, planting pumpkins at the edge of the patch to make the terrain hard to cross, and even fastening paper bags over the ears — a real last resort — help. Deer damage often can be controlled by spreading woven fencing — chicken wire or other — flat on the ground at the edge of the patch that they would have to cross to get to the corn. They will not like it when their hooves get caught in the mesh. They — and other animals — also respect lion manure scattered around the garden (see your local zoo or traveling circus).

Birds

Crows and redwing blackbirds like to work their way down a row of just-sprouted corn, plucking up each tender spear to get at the buried seed kernel. They lose interest when sprouts grow taller than about three inches, so you can transplant corn seedlings into your row if you have a serious crow problem. Just start your seedlings early in peat pots. Or, stretch a random netting of string, three to five inches above the ground, across the patch to foil the birds. Seeds that are painted with tar are usually safe from crows, but they are so sticky that they are hard to plant in any quantity. Avoid leaving unburied seed scattered in the corn patch when planting. Birds are often attracted to this exposed seed and then go scratching for more. You might also try hanging pieces of fur around the garden — and putting up a scarecrow (which should be changed from time to time; these birds are smart!).

A string stretched randomly across the corn patch like a "cat's cradle"
will discourage birds from plucking your tender corn sprouts. A scare-
crow will keep birds away, too.

Harvesting

Sweet Corn. Sweet corn is at its tender, sweet, juicy best for only a
few days. That moment of perfection occurs about eighteen to
twenty days after the silks have been pollinated. This is the milk
stage, when the kernels contain as much moisture as they will ever
have. The juice in the kernels is milky, and the usual test of readiness
is to puncture a kernel with your fingernail to see if milky juice

spurts out. If you are too early, the juice will be watery. Too late (about twenty-eight days beyond pollination), and the kernels will turn doughy as the moisture recedes and sugar turns to starch. Here are some other signs of ready-to-pick corn.

- Dark green husks. If the husks are yellow, the corn is probably old and tough.
- Brown, but not brittle, silks.
- Well-filled ears. It is important to know your variety here, though. The varieties Wonderful and Candystick, for example, have very slim ears and can be ready before you realize it, if you are waiting for those ears to fill out.

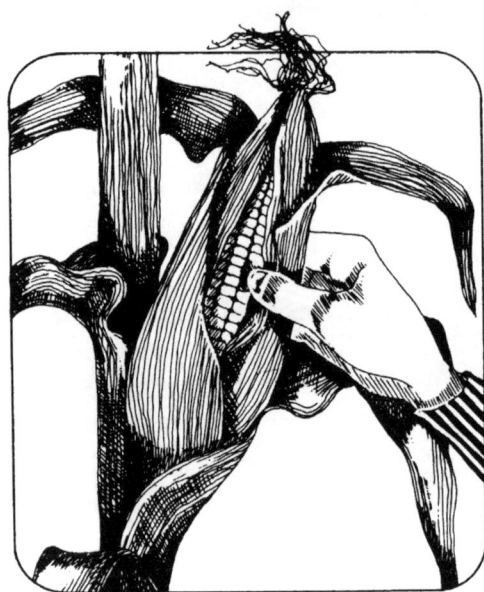

Sweet corn is at its harvesting best at the "milk stage." If you puncture a kernel at this time, the juice in the kernel will be milky. If you are too early, the juice will be watery.

Your harvest dates may vary somewhat from the maturity dates you see listed in the seed catalogues. Eighty-day corn, for example, is not always ready exactly eighty days from planting. Prevailing temperatures and rainfall, soil quality, and even microclimate conditions in your own garden can easily add several days to the growing period.

Sweet corn should be eaten or processed as soon as possible after picking, because in standard varieties the sugar starts to change to starch as soon as the ear is plucked. Cooling the corn helps to slow this process. If you do not cook the corn immediately after picking, leave the husks on the ears, and store the corn in plastic bags in the refrigerator, If you have never tried raw sweet corn, treat yourself to an ear or two while husking.

Popcorn. Leave the ears on the stalk until the kernels are hard and dry. They need to develop a hard skin to contain the small burst of steam that builds up inside each kernel when it is heated, or the pop will be just a sputter, and the interior contents will not puff up. Pull off the ears and peel back the husks. It is a good idea to hang the ears in a dry, airy place to cure for a week or two after picking. You can tie several ears together by the husk and drape them over a nail, or you can string a wire and hang a series of ears along its length. Or you can husk the ears and store them in open baskets.

Flint Corn. Most people grow flint corn for grinding, so it must be good and dry when picked. Kernels should appear glazed, and cannot be indented by a thumbnail. Damp corn will mold in storage and gum up the grinder. If the weather is dry enough, leave the ears on the stalk well into the fall, until they are thoroughly dry. If you must pick the ears before they are fully dry, complete the drying process in an oven, over a pilot light, or near a wood stove.

Dent Corn. Like flint corn, dent corn should dry completely on the stalk before picking. The whole plant should be brown and dry and each kernel well indented before the corn is picked for storage. For immediate use as animal feed in early fall, ears can be hand-picked and husked as needed. When the corn is good and dry it will be easier to remove from the husk and easier to grind for household use. Dent corn can be left in the field and gathered as needed, but in areas where snow is common, it is best to harvest the ears while you can still get a cart into the field. Heavy snow or roaming deer can do away with a lot of ear corn left on the stalk over winter.

How to Save Your Own Corn Seed

Saving your own seed is not only a smart economy, it is also a way to develop a strain of vegetables especially well suited to your particular growing conditions.

Since seed saved from hybrid corn does not come true, you will have to confine your seed-savings efforts to open-pollinated corn varieties, such as Golden Bantam, White Evergreen, and Country Gentleman. Separate your seed corn from other tasseling corn by 500 to 1300 feet.

Begin by deciding which plants and ears you want to use for your seed stock. Look for earliness, flavor, ear size, disease-resistance, and insect-resistance. When you have selected the best parent plants, mark them well to be sure that no one picks and eats those ears! Use a stake, a red strip of fabric, or a net bag slipped over the ear after it has been pollinated. (Browning of silks indicates that pollination has taken place.)

If they are to germinate well in the next growing season, seeds must be allowed to ripen fully on the plant. So leaves your seed ears on the stalk for about a month past the time wen they would have been just right for eating. Frost will not hurt well-dried seed corn. It is a good idea to hang the seed ears, with husks pulled back, in a well-ventilated spot to dry some more after picking. Even if you need only a handful of seed corn, save and mix the seed from several plants so your corn will not become inbred.

You can store your seed corn on the ear in cans, or shelled in jars or cans. Keep it as dry as possible and in a cool, dark place. Heat and moisture use up stored nutrients in the seed and reduce germination rates. If kept completely dry and quite cool, corn seed can remain usable for as long as five years. To be on the safe side, most gardeners count on planting seed corn no older than two to three years.

Dry flint, dent, seed corn, and popcorn by suspending the ears from a wire or rope, with their husks pulled back.

Storage

In the frontier settlements of nineteenth-century America, having enough food meant having sufficient corn. Today we enjoy a much wider variety of foods, but corn can be a source of a great many delicious dishes, from steamed sweet corn to cornmeal mush, crackers, muffins, corn pudding, popcorn, tortillas, chapaties, corn chowder, succotash, stewed dried corn, and others. We have more ways to keep corn now, too. Here is how you can store your corn crop.

Freezing

Freezing is strictly for sweet corn. For whole-kernel corn, scald ears of husked corn in boiling water to cover for four minutes. Chill, then cut the kernels off with a sharp knife. Then package and freeze. To freeze whole ears, scald small ears for seven minutes, medium ones for nine minutes, and large ears for eleven minutes. Then, drain, chill, and package.

Canning

Again, canning is a method almost exclusively used for sweet corn, although in a pinch one could can green (immature) dent corn. Corn is a low-acid vegetable and must be canned under pressure. Never use hot water bath or open-kettle methods.

To can whole-kernel corn, cut the kernels from the cob and put into clean jars, leaving a one-inch space at the top. Do not pack the corn tightly. Pour boiling water over the corn to cover it, leaving a one-inch head space. Put caps on jars. Process pints for fifty-five minutes, quarts for one hour and twenty-five minutes, at ten pounds of pressure.

Ear Storage

Popcorn can be stored on the ear in discarded nylon stockings or mesh bags, or in cans; or it can be shelled and kept in jars in a cool, dry place. To shell popcorn, rub two ears together or put an old sock over your hand and rub the kernels off. One-half cup of popcorn

Canning Whole-Kernel Corn

1. *Cut the kernels from the cob.*
2. *Pour into clean jars, leaving a one-inch head space.*
3. *Pour boiling water over the corn to cover it, leaving a one-inch head space. Cap the jars.*
4. *Process at ten pounds of pressure, pints for fifty-five minutes, quarts for one hour and twenty-five minutes.*
5. *Test the seals.*

kernels should yield about four quarts of popped corn. If you try some and it does not pop well, it is probably too dry. Add one tablespoon of water to three cups of kernels and shake; then cap and leave for one to two days. Kernels should not be kept wet, though, or they will mold. Some folks say that popcorn pops better when it has been refrigerated.

Flint and dent corn can be stored on the ear in bins or cribs, or in cans; or it can be shelled by hand or hand-cranked machine and kept in cans, as long as it is well dried before being covered.

Corn Varieties to Grow

The following list of corn varieties is just a sampling of the many delicious strains available. Each seed house carries a different assortment, and there are many more good corn varieties out there to try. These are some of the most widely grown selections. Perhaps some of them will become your family favorites.

SWEET CORN

Early
Polarvee: 53 days
Spancross: 60 days
Earlivee: 60 days
Sprite: 68 days bicolor

Midseason
Quicksilver: 72 days; 3 weeks earlier than Silver Queen
Wonderful: 82 days
Butter and Sugar: 80–84 days
Golden Bantam: 83 days open-pollinated
Country Gentleman: 92 days open-pollinated
Stowell's Evergreen: 99 days; an oldie grown since the mid-1800s
Black Mexican: 86 days open-pollinated, kernels blue-black at maturity

Late
Jubilee: 87 days
Honey Cross: 87 days wilt resistant, tight husks
Silver Queen: 92 days large, late, and delicious
Honey and Cream: 92 days bicolor

Super-Sweet Varieties
Earliglow: 70 days everlasting heritage
Kandy Korn: 89 days everlasting heritage
Florida Staysweet: 84 days less trouble with germination and seedling vigor; needs isolation
Iochief: 89 days; needs isolation
Illinichief: 85 days needs isolation

Wilt-Resistant Varieties
Honey Cross: 87 days
Golden Cross Bantam: 75 days; hybrid
Comanche: 70 days
Bellringer: 79 days
Comet* white Bi-Queen: 92 days bicolor

POPCORN

Rhodes Yellow Pop: 118 days; 2–3 ears per stalk
White Hull-less Hybrid*
Strawberry: 105–110 days open-pollinated
White Cloud: 95 days hybrid
South American* 2–3 ears, golden kernels

FLINT CORN

Longfellow: 117 days
Garland: 110 days
Indian Flint: 105 days multicolored ears, but meal is white
Rainbow Flint* multicolored ears, but meal is white
Nichols Hominy Corn: 78 days

FLOUR CORN

Mandan Bride: 98 days multicolored ears

DENT CORN

Seneca Hybrid: 103 days
Reid's Yellow Dent: 85 days

*No maturation time available

Seed Sources

Fedco Seeds
207-426-9900
www.fedcoseeds.com

Heirloom Seeds
724-663-5356
www.heirloomseeds.com

Johnny's Selected Seeds
877-564-6997
www.johnnyseeds.com

Nichols Garden Nursery
800-422-3985
www.nicholsgardennursery.com

Park Seed
800-845-3369
http://parkseed.com

Peaceful Valley Farm Supply, Inc.
888-784-1722
www.groworganic.com

Pinetree Garden Seeds
207-926-3400
www.superseeds.com

Seed Savers Exchange
563-382-5990
www.seedsavers.org

Seeds of Change
888-762-7333
edsofchange.com

e Burpee & Co.
·1447
pee.com